Amazon Echo
User Guide To Understand Amazon Echo Fast!

© **Copyright 2016 Ted Lebowski - All rights reserved.**

This document is geared towards providing exact and reliable information in regards to the topic and issue covered. The publication is sold with the idea that the publisher is not required to render accounting, officially permitted, or otherwise, qualified services. If advice is necessary, legal or professional, a practiced individual in the profession should be ordered.

- From a Declaration of Principles which was accepted and approved equally by a Committee of the American Bar Association and a Committee of Publishers and Associations.

In no way is it legal to reproduce, duplicate, or transmit any part of this document in either electronic means or in printed format. Recording of this publication is strictly prohibited and any storage of this document is not allowed unless with written permission from the publisher. All rights reserved.

The information provided herein is stated to be truthful and consistent, in that any liability, in terms of inattention or otherwise, by any usage or abuse of any policies, processes, or directions contained within is the solitary and utter responsibility of the recipient reader. Under no circumstances will any legal responsibility or blame be held against the publisher for any reparation, damages, or monetary loss due to the information herein, either directly or indirectly.

Respective authors own all copyrights not held by the publisher.

The information herein is offered for informational purposes solely, and is universal as so. The presentation of the information is without contract or any type of guarantee assurance.

The trademarks that are used are without any consent, and the publication of the trademark is without permission or backing by the trademark owner. All trademarks and brands within this book are for clarifying purposes only and are the owned by the owners themselves, not affiliated with this document.

Table of Content

What is Amazon Echo?... 1

Setting up Your Amazon Echo... 3

How to Restart Your Echo Device... 5

Hardware Basics... 6

The Echo Remote... 7

Differences between Echo, Echo Dot, and Amazon Tap 9

Using Alexa on Other Alexa-Enabled Devices............................... 13

Commands for Alexa... 15

How to Listen To Music.. 17

Reading Kindle Books with Alexa 20

Linking Third-Party Music Services... 23

Connecting Smart-Home Devices to Alexa 25

Supported Smart-Home Devices 28

Removing a Smart Home Device 29

Creating Smart Home Groups 30

Undiscoverable Smart Home Device... 32

Commands to Try 34

Weather Updates 36

Changing Your Device's Location 37

Requesting Music 38

When Alexa Doesn't Understand You 40

Voice Training 42

Bluetooth Issues 43

When the Alexa App Doesn't Work... 46

Streaming Issues 49

Software Updates 51

Kindle/Fire Software Updates 53

What is Amazon Echo?

Amazon Echo is a wireless speaker you can control with your voice, no hands necessary. You can ask it to play music, gather news, weather, sports, etc. Amazon Echo can be heard from across the room or over the music, as it consists of seven different microphones and beam forming technology. The Amazon Echo device is a smudge over three inches wide, nine inches tall and the speaker is strongly arrayed. The expertly tuned speaker can fill up any room with its 360° immersive sound.

Whenever you want to use the device, just say the wake word "Alexa" and it responds immediately. If you own more than one Echo or Echo Dot device, you can choose to set a different wake word for each. For instance, you can set wake words like "Amazon", "Alexa" or "Echo". Whenever you decide to use Echo, you can easily shout whichever wake word you decided to set for your device and it will respond immediately for your command.

Amazon Echo provides hands-free voice control for various applications like Pandora, iHeartRadio, Spotify, TuneIn or Prime Music. Amazon Echo is also available for Bluetooth so you can play music from any application like iTunes directly from your phone or tablet. It's even better if you use Amazon Echo frequently because the more you do use it, the more intelligent it becomes as it absorbs and get used to your speech patterns, vocabulary and preference.

As far as the physique of the device, there are three things that you can fiddle with: the microphone button (which you can use to switch on or off the listening feature), the action button (which summons Alexa when it's tapped without using the wake word and/or brings up a manual step when held), and the volume ring. Also, when you plug in your Amazon Echo device, you'll want to grab the top of the speaker and adjust the volume. To lower the volume, you would have to turn it counter clockwise. Now that we told you information about the physiques of the Amazon Echo, let's move on to other details like how to set up your Echo.

Setting up Your Amazon Echo

The first thing you would have to do is download the Alexa app and sign in so you can avoid Alexa from taking up your time just from filling you in about instructions. You can manage things like your music, shopping lists, alarms and more just with the Alexa app. You can download the Alexa app for free, which is available on select phones and tablets with:

iOS 7.0 or higher
Android 4.0 or higher
Fire OS 2.0 or higher

You can download the Alexa app either on the app store or google play store by searching for "Alexa app" and then select and download it. Also, keep in mind that the 1st and 2nd generation of the Kindle tablets is **NOT** compatible with the Alexa app.

The first thing you want to do to set up your Echo is obviously turn it on. Then, the next thing you want to do is plug in the power adapter that came with the Echo into the speaker and into an outlet. After you do this, the speaker's ring will shift from blue to orange. Echo will officially greet you once the ring turns orange. Next, connect the device to a wireless network (Wi-Fi) to activate it and get it ready for internet use. You can also follow the guided instructions in the app to connect Echo to a Wi-Fi network if you're having trouble.

To begin using your device, say the default wake word "Alexa" then begin speaking to it/giving it commands. You can also change the wake word just by going to Settings > Select your device > then select the certain wake word you would like to use.

If your speaker won't connect to your Wi-Fi network for some reason, you can easily restart it by unplugging it then plugging the device into an outlet. If that doesn't make a difference, you can reset your Echo to its factory settings, which will allow you to set it up all over again. To learn more, you can skip to the next section **"How to Restart Your Echo Device"**.

How to Restart Your Echo Device

Since your speaker won't connect to your Wi-Fi network, you can just restart the device to see if it resolves your problem. As mentioned before, you can choose to unplug the power cord from the back of your Echo device or from the wall outlet. Then, plug the power adapter back in.

To restart your Echo device:

1. Press and hold the reset button at the base of your device for five seconds using a small tool like a bobby pin or paper clip. Use a paper clip (or similar tool) to press and hold the reset button for five seconds. After you have done so, the light ring on your device will turn orange then blue.
    2. The light ring will turn on and off again then after that, it will turn orange. This means that your device went back to setup mode.
    3. In the Alexa app, connect your device to a Wi-Fi network and register it to your Amazon account.

Hardware Basics

Here's a list of all the hardware features of your Echo device and what it does:

Microphone Button – Turns off all the microphones. When the light ring is red, that's how you know you have successfully turned it off. To turn it back on, just simply press it again.

Light Ring - Notifies you what the device's current status is.

Volume ring – Increases/decreases the device's volume. To increase the volume, turn it clockwise. To decrease it, turn it counterclockwise. The light ring will display the volume level in white as you increase/decrease it.

Action Button - The action button can be used to turn on/off a timer or alarm, wake your device up or to enable the Wi-Fi setup after holding it for five seconds.

Power LED - The Power LED lets you know what the current Wi-Fi status is on the device. If it's a white light, your device is connected to a Wi-Fi network. If it's orange, your device is **not** connected to a network. If it's orange and blinking, it is connected to a network but cannot access Alexa.

The Echo Remote

What's amazing about the Amazon Echo Device is that it includes a voice-enabled, wireless, battery powered Echo Remote. You can control audio playback on your Echo device just because of the directional track pad that is featured with it. Keep in mind that you can only pair/use one remote per Echo device and that the Echo Remote is sold separately. Read to the next section for instructions on how to set it up. Also, it's compatible with Echo and Echo Dot **only**.

The first thing you want to do is install the batteries into the remote. Insert the two AAA batteries that was included with your Echo Remote inside of it and put the battery door back on. You will be able to pair the remote once you install the batteries. To pair the remote, open the Alexa app > open the navigation menu > select Settings. It will prompt you to select your Echo device then from that point, select Pair Remote.

Note: If you see **"Forget Remote"** on the app, this means that another remote is already paired with your Echo device. This means that a remote is already paired with your device. If you're replacing the previous remote, select Forget Remote before pairing the new one with your device.

To search for your remote, press and hold the play/pause button for five seconds before releasing it. It will search for your remote and connect to it within

40 seconds once it has been located. Alexa will tell you're your remote has been paired.

If the Amazon Echo Remote doesn't pair or work with your Echo device, here are some troubleshooting in attempt to get it to work again:

·Try installing new batteries
It's highly possible for batteries to lose their energy if they haven't been used for a long time. In circumstances like these, you would have to purchase brand new AAA batteries and install it. Make sure they're installed correctly when you put them inside the Remote.

·Try pairing the remote again

Differences between Echo, Echo Dot, and Amazon Tap

Echo, Amazon Tap, and Echo Dot are all different when it comes to their hardware. What they do have in common is that they all access the Alexa Voice Service. I'll explain to you how it's different from one another:

Portable:

Echo: No
Amazon Tap: Yes
Echo Dot: No

Power:

Echo: Power adapter
Amazon Tap: Non-removable, rechargeable battery
Echo Dot: Power adapter

Wi-Fi:

Echo: 802.11a/b/g/n, dual-band (2.4 GHz and 5 GHz)
Amazon Tap: 802.11b/g/n, single-band (2.4 GHz)
Echo Dot: 802.11a/b/g/n, dual-band (2.4 GHz and 5 GHz)

Alexa Activation:

Echo: Wake word, Action button
Amazon Tap: Microphone (Talk) button
Echo Dot: Wake word, Action button

Speakers:

Echo: Mono
Amazon Tap: Stereo
Echo Dot: Stereo

Dolby Audio:

Echo: No
Amazon Tap: Yes
Echo Dot: No

Buttons and Lights:

Echo: Mute button, Light ring, Action button
Amazon Tap: Front light indicator, Microphone button, Power button, Wi-Fi/Bluetooth button, dedicated playback buttons
Echo Dot: Light ring, Action button, Mute button

Bluetooth Audio Input:

Echo: Yes
Amazon Tap: Yes
Echo Dot: Yes

Bluetooth Audio Output:
Echo: No
Amazon Tap: No
Echo Dot: Yes

AUX Audio Input:

Echo: No
Amazon Tap: Yes
Echo Dot: No

AUX Audio Output:

Echo: No
Amazon Tap: No
Echo Dot: Yes

Compatible with Voice Remote for Amazon Echo:

Echo: Yes
Amazon Tap: No
Echo Dot: Yes

Accessories:

Echo: Voice Remote for Amazon Echo (sold separately)
Amazon Tap: Charging Cradle (included), Amazon Tap Sling (sold separately)
Echo Dot: 3.5mm audio cable (included), Voice Remote for Amazon Echo (sold separately)

Media Storage:

Echo: No
Amazon Tap: No
Echo Dot: No

Alexa App:

Echo: Yes
Amazon Tap: Yes
Echo Dot: Yes

Using Alexa on Other Alexa-Enabled Devices

The Alexa Voice Service is the service that powers Alexa-enabled products such as the Amazon Echo and other products that Amazon do not manufacture.

To use Alexa on other Alexa-enabled products:

- Set up your product and connect it to the Internet. Refer to the product companion app or website for more information.
- Sign into your Amazon account in the product companion app/website in order to enable Alexa.
- To manage Alexa settings, download the Amazon Alexa App. You can also manage your Music & Media, Smart Home and tons more in the app.
- Open the Alexa app and make an Alexa request. Go to Things to try for examples.

Note: Some Alexa-Enabled products doesn't support all of the Alexa features. You can see the settings/features that aren't supported by your other Alexa-enabled products in the Alexa app.

About the Alexa App

With the Alexa app on Fire OS, Android, iOS, and desktop browsers, you can manage your Alexa settings. The Alexa app allows you to do the following:

Configure Alexa-enabled product settings
Access your dialog history

View your Shopping and To-do lists
Configure timers and alarms
Discover and set up smart home devices
Discover and enable third-party skills
Check out new "Things to Try"

Need help?
Contact the device manufacturer for any of these issues:

Setting up your device
Internet connection problems
Alexa activation (e.g. Wake Word)
Supported Alexa features
Device-specific skills

Commands for Alexa

You can control Alexa just by the sound of your voice. You can even do this as you listen to audiobooks, music and more once you have paired your mobile or tablet device with your Alexa device.

Note: Voice control is currently not available for Mac OS X devices (such as MacBook Air) or on other Alexa-enabled products.

If you say the word "Connect", your Alexa device will begin to search for your paired mobile device and connect to it immediately. To play a song, open a music app on your mobile device and select the song you want to play. It will begin playing from your Alexa device.

Use these voice commands to control playback on your Alexa device:

- Play
- Pause
- Previous
- Next
- Stop
- Resume
- Restart

Alexa will pause playback on your mobile device and disconnect from it if you request songs, albums, artists or playlists while listening to music from your mobile device. When this happens, music from your Amazon

Music library will start playing instead. To simply play music over Bluetooth again, press play on your mobile device and tell Alexa, "Connect."

How to Listen To Music

Alexa can be used to stream music, podcasts, audiobooks, and more from popular streaming services or from your phone/tablet over Bluetooth. You can also upload your personal music collection from music apps like iTunes, Google Play Music, and more to your music library on Amazon and then play it through your Alexa device.

The list below is a growing number of free and subscription-based streaming services that are supported on Amazon devices by Alexa:

- Prime Music
- Pandora
- Audible
- Amazon Music
- iHeartRadio
- Spotify Premium*
- TuneIn

You can ask Alexa to stream music, audiobooks, live radio, podcasts, and more directly from these services through your Alexa device. You automatically have access to any of the music available in your Amazon Music library after you register an Alexa device to your Amazon account. Amazon Prime members can also enjoy playlists, stations, and over a million free songs with Prime Music. Other services may require you to link an existing account or subscription to your Amazon account in the Alexa app. You can find out more information about this on the **"Linking Third-Party Music Services"** section.

Note: Spotify Premium on Alexa supports Spotify Connect, which allows you to play music on your Alexa device and use the Spotify app on your phone or tablet as a remote.

Upload Your Music

You can actually play your own music on your Alexa device. To play your personal music from apps like iTunes, Google Play, and more on your Alexa device, you can use Amazon Music for PC or Mac to upload your collection to your music library on Amazon directly from your computer. Alexa will be able to play your music and control playback with voice commands once you have uploaded your music to your music library.

Without an Amazon Music Subscription, you're allowed to upload up to 250 songs to your music library. With an Amazon Music Subscription, you're allowed to upload up to 250,000 songs. Purchases from the Digital Music Store on Amazon is not considered part of the 250 song limit.

Stream Music and Media over Bluetooth

The Advanced Audio Distribution Profile (also known as A2DP) are supported by Echo, Echo Dot, and Amazon Tap to stream audio over a Bluetooth connection. The Audio/Video Remote Control Profile (AVRCP) is for voice control of connected mobile devices. Echo Dot also supports A2DP SRC, which gives you the privilege to stream audio from Echo Dot to a Bluetooth speaker.

Reading Kindle Books with Alexa

You can ask Alexa to read eligible Kindle books in your library if they're by Alexa. Your Kindle books can be read by Alexa with the same text-to-speech technology used for Wikipedia articles, news articles, and calendar events.

Alexa can only read eligible books that are:

- Purchased from the Kindle Store
- Borrowed from Kindle Owners' Lending Library
- Borrowed from Kindle Unlimited
- Shared with you in your Family Library
- To find eligible books in the Alexa app, open the left navigation panel and then select Music & Books > Kindle Books > Books Alexa can read.

Note: Comic/graphic novels and narration speed control content/features are not supported by Alexa.

Alexa being able to read your books for you is an

amazing feature because when you ask Alexa to read your Kindle book, she picks up right from where you left off in the book, even if it's from another compatible Amazon device or reading app. In order for Alexa to read for you, the Kindle books must be eligible for Text-to-Speech (an experimental reading technology that allows supported Amazon devices to read Kindle books aloud).

Tip: Select the Now Playing bar in the Alexa app when Alexa reads your Kindle book and then select Queue to go to different chapters in a Kindle book. You can then choose a chapter from the list.

To listen to a kindle book, you can say these commands:

- "Read my Kindle book"
- "(wake word), read (book title)."
- "Read my book (title)"
- "Play the Kindle book (title)"
- "Read (title)"

To pause a Kindle book, you can easily say:

- "Pause."

- "Stop."

To continue listening to your Kindle book, you can say:

- "Play."
- "Resume."

- To go to the next or previous paragraph, say:
- "Skip back."
- "Skip ahead."
- "Go back."
- "Go forward."
- "Next."
- "Previous."

Note: Whenever you skip ahead/back, it will do so in the text by 30 seconds.

Linking Third-Party Music Services

To listen to other streaming music services, you may need to link an existing account from the service to Alexa in the Alexa app. These are the music services that requires you to link your account to Alexa:

Spotify Premium
iHeartRadio
Pandora

Be aware that some of the music services are not available on other Alexa-enabled products (manufactured by other companies) for various reasons. Check with the product manufacturer for more information.

To link the service to your Alexa device:

- Open the left navigation panel in the Alexa app
- Select Music & Books, and then select the music service of your choice.
- Select Link account to Alexa and then a sign-in page will appear on the screen.
- Sign in using the email address and password for the service. Please don't get confused about the sign-in, it is not the same as the email address and password for your Amazon account.
- If you receive an error message when you try to link your account to Alexa, reset your username and password for the streaming service and then

try to link your account to Alexa again.

Note: Use of streaming music services may be subject to additional terms applicable to the respective service. You can unlink Alexa from a streaming music service at any time by selecting Unlink account from Alexa in the Alexa app.

Connecting Smart-Home Devices to Alexa

You can enable the skill to connect your smart home devices to Alexa after you have set it up to one. You may want to check **"Supported Smart-Home Devices"** to see if your smart home device is compatible for Alexa.

Before you start, download the manufacturer's companion app for your smart home device on your mobile device. The app is used to set up the smart home's device on the same internet network connection as your Alexa device. Be sure to download and install the latest software updates for your device so you won't be out of date.

To connect your smart home device to Alexa:

- Select "Skills" after opening the Alexa app.
- Then select "Categories" and choose "Smart Home".
- Search for the keywords of the skill for your smart home device and then select "Enable Skill" once you have found it.
- It may ask you to sign in with your third-party information but if it does, sign in and follow the instructions to complete the setup.
- Select "Discover devices" or say "Discover my devices" once you have enabled the skill.
- You should now be able to control the device just by using your voice. The word "Unreachable" will

appear next to the device if Alexa didn't discover it.

If you restarted your smart home device, it may take a couple of minutes for Alexa to rediscover it. If Alexa still fails to discover your device, check your smart home device on the companion app to be sure it's connected to the same internet network as your Alexa device.

To search for available smart home skills:
- Open the Alexa app and select "Skills".
- Select "Categories".
- Choose "Smart Home".
- Brightness Control

You can actually set the brightness of compatible light bulbs with some of the smart home skills, which is pretty cool. You can also control things like switching on/off a lamp, turning on/off a fan, or increasing/decreasing room temperatures. Note that some of the products work directly with Alexa and other smart home ecosystems require a compatible hub. The performance of these devices varies when you attempt to control them with your Alexa device.

After you enabled the smart home skill of your choice and added your smart home device, you can start saying simple commands to control it with Alexa. To learn more about the devices you can control with Alexa, go to **"Supported Smart Home Devices"**.

For some smart home skills, you actually need to say

"**Open (skill name)**" before making your request.

Say this to do this:

- To turn on / off your smart home device:
  "Turn on / off (smart home device / group name)."
- To set the brightness of compatible lights
  "Set (smart home device / group name) to (##) %."
- To control a thermostat
  "Set (smart home device / group name) temperature to (##) degrees."
  "(Increase / decrease) the (smart home device / group name) temperature."
- To change your fan's speed
  "Set my bedroom fan to (##) %."

Some of the smart home skills may support other device actions. For more information go to the smart home skill in the Alexa app.

Supported Smart-Home Devices

Not all smart home devices are going to be compatible with Alexa. To see if your device is compatible with Alexa, check with the list of devices below:

- Samsung SmartThings
- Philips Hue
- Wemo
- Insteon
- Wink
- Lutron

Removing a Smart Home Device

At some point, you may want to remove your smart home device(s) from your Alexa device so you can either add another one or just simply forget that device. To remove a smart home device, you can choose the "forget" option in the Alexa app.

Note that turning off your Alexa device doesn't remove any connected device. It will still recognize the smart home device once it's plugged in again. To actually remove the smart home device, you must:

- Open the left navigation panel and then select "Smart Home" in the Alexa app.
- Under "Devices", select Forget for the device(s) you wish to remove.
- You can also forget all of the devices at once, just disable the skill in "Skills" using the Alexa app.

Creating Smart Home Groups

To control multiple smart home devices at one time, you can easily create a group in the Alexa app. You'll be able to control all the devices with simple commands like "Turn off bedroom lights" or "Dim bedroom lights to 50%" and many more. For best results, add categorized devices to a group. For example, keep thermostats separated from lights and switches.

To create a smart home device group:
- Select "Smart Home" located in the left navigation panel on the Alexa app.
- Select "Create Group" under the "Group" section.
- Enter the name of the group into the text field. It's recommended to give your group a recognizable name for Alexa to identify. Use names that have at least two to three syllables. If you want to name multiple groups, be sure to give each of them a different name. For example, if you have devices in your kitchen and basement, you could simply name them "Kitchen" and "Basement".
- Select the smart home device(s) you want to add to the group, and then select add.

To edit a smart home device group:
- Select your smart home group.
- Make changes to your group
- To edit the name, select the text field and then update the existing name.
- To add or remove smart home devices, select the checkboxes next to each device.

Undiscoverable Smart Home Device

It's possible that your Smart Home can't be discovered by Alexa and it comes with various reasons why. Here are some tips if Alexa device doesn't discover your smart home device (such as Wink or SmartThings):

- First, check to see if your smart home device is compatible with your Alexa device. To see if your device is compatible, refer back to **"Supported Smart Home Devices"** for the list of compatible devices.
- Download the companion app for your smart home device in the app store, and then set up your device.
- Restart both the Alexa device and smart home device.
- Disable then enable the smart home skill in the Alexa app.
- Download and install any software updates for your devices since out of date updates can strongly affect it.
- Connect to the same Wi-Fi network with both your Alexa and smart home device. If your Alexa device is not on the same Wi-Fi network, use the Alexa app to change to the network

To change the network:
- Open the Alexa app.
- Open the left navigation panel.
- Select Settings.
- Select your Alexa device

- Select Update Wi-Fi. Follow the instructions in the app to update your Wi-Fi information.
- Alexa and smart home devices work best on personal Wi-Fi networks. Wi-Fi networks at school or work may not allow unrecognized devices to connect.

Some smart home devices can only connect to the 2.4 GHz Wi-Fi band.

Turn on SSDP / UPnP on your router
Use your computer to update your router settings. Contact your router manufacturer for assistance.

Rename your smart home device
Check that the group name you've assigned to the smart home device in the Alexa app can easily be understood by Alexa device.

Discover your devices again

If you say "Discover my devices", Alexa will start searching for your devices. When discovery is complete, Alexa will tell you what she has found or if she didn't find any devices.

If you choose to delete the group, simply select Delete.

Commands to Try

You can do anything ranging from finding news, restaurants, activities to asking questions, control music with your voice, etc. When it comes to questions, you can ask Alexa any type of question. For example:

- "How old is Ashton Kutcher?"

- "What's the capital of Florida?"

- "What year did ACDC form as a band?"

- "How many degrees is it outside?"

To control music with your voice, you can say these:

- "Alexa, what's playing?"

- "Alexa, turn it up."

- "Alexa, mute."

- "Alexa, stop the music."

- "Alexa, pause."

- "Alexa, resume."

- "Alexa, next song."

- "Alexa, stop playing music in 30 minutes."

If you're listening to music from a third-party service, you can also say these:

- "Alexa, buy this song / album" (while a music sample or music on a radio station is playing).

- "Alexa, add this song" (while Prime Music is playing).

- "Alexa, I like this song" (when a song or track from Pandora / iHeartRadio / Prime Stations is playing).

- "Alexa, thumbs down" (when a song or track from Pandora / iHeartRadio / Prime Stations is playing).

Weather Updates

You can actually ask Alexa about local, national, and international weather forecasts. She can tell you how the weather is currently and even the weather forecast in other states and cities. To get started, add your address in the Alexa app. For more information, go to **"Changing Your Device's Location"**.

Alexa will respond to questions like these:

- Alexa, what's the weather?

- Alexa, will it rain tomorrow?

- Alexa, is it going to rain on Wednesday?

- Alexa, what's the weather in Miami?

- Alexa, what will the weather in Kansas City be like tomorrow?

- Alexa, what's the weather in North Dakota?

- Alexa, what's the weather in Las Cruces, New Mexico?

- Alexa, what's the extended forecast for Salt Lake City, Utah?

- Alexa, what will the weather be like in West Palm Beach on Sunday?

Changing Your Device's Location

Whether you constantly move to different locations or need to update your location, you can actually modify it in your Alexa device. To change the location:

- Select **"Settings"** under the left navigation panel in the Alexa app.
- Select your device.
- In the **"Device location"** section, select **"Edit"**.
- Enter the complete address, including the street name, city, state, and ZIP code.

If you have multiple Alexa devices on your Amazon account, make sure to update the location for each device.

Requesting Music

When you buy music from the Digital Music Store, you can immediately play it from your music library. You can also import songs to your Amazon Music library and play it on your Alexa device. Eligible Prime members can also listen to Prime Music.

To request music, you can tell Alexa commands like these:

- Alexa, play the song I just bought.

- Alexa, play some music.

- Alexa, play music by Guns N Roses.

- Alexa, play "Radioactive" by Imagine Dragons.

- Alexa, shuffle my rock playlist.

- Alexa, play Skrillex.

- Alexa, play new music by Future.

- Alexa, shuffle my new music.

- Alexa, play the album "Toxicity".

- Alexa, play the song "Down With The Sickness"

- Alexa, play some dubstep.

When Alexa Doesn't Understand You

There are moments where Alexa won't understand you. It can either be because of your speech, a weak Wi-Fi connection or the location. Here are some tips if Alexa does not understand your requests:

In order for Alexa to work, you need an active Wi-Fi connection. An active Wi-Fi connection allows Alexa to stream music and other media and answer your questions or process your requests. To make sure Alexa can do those things:

- Leave your Alexa device in an ideal location
- Walls and other objects can actually cause interference for the device so make sure that it is at least eight inches away from these obstacles.
- Move your Alexa to a higher location if it's on or near the floor.
- Make sure there is no background noise as you give commands to Alexa, since it can interfere and give her a hard time to understand you. Also, speak naturally and clearly to Alexa.
- Be specific.
- Repeat your question or request if she doesn't understand.
- Rephrase your question or make it less general. For example, if you want to know something about a certain city, include the state as well since there can be duplicates of a city name. (i.e.; "Miami, Florida").
- Check the Alexa app to see what Alexa heard. On the home screen, select **"More"** at the bottom of

the interaction card. You can read what Alexa heard, listen to the request, or provide feedback.
- (Optional) Use Voice Training

Voice Training can be used for both Amazon Tap and Echo devices. It helps Alexa understand your speech patterns. As you use Voice Training, the Alexa app will show the 25 phrases you use/say with your device. If you want to read about more information about Voice Training, head on to the next section.

Voice Training

You can use Voice Training in the Alexa app to improve the speech recognition on your Alexa device when she doesn't understand you. Voice Training is only allowed for use with Echo and Amazon Tap devices. During a Voice Training session, say each of the 25 different phrases to your device. Your device processes every phrase you say, even if you don't finish a session. For best results:

- Speak normally to your Alexa device.
- Sit or stand where you normally speak to your Alexa device.
- Don't use a voice remote during voice training to avoid interruptions.

To start a Voice Training session:
- Start the Alexa app.
- Open the left navigation panel, and then select **"Settings"**.
- Select **"Voice Training"**.
- Select **"Start Session"**.
- Speak the phrase in the Alexa app, and then select Next.
- To repeat a phrase, select Pause, and then Repeat Phrase.
- When you reach the end of your session, select Complete.
- If you need to end your session at any time, select Pause, and then End session.

Bluetooth Issues

If your Alexa device won't connect over Bluetooth, here are some tips for your mobile device (phone or tablet) to help you troubleshoot the issue. This will work for Echo Dot, Echo and Amazon Tap.

The three devices (Echo, Echo Dot and Amazon Tap) support these Bluetooth profiles:

- Advanced Audio Distribution Profile (A2DP SNK)

This profile allows you to stream audio from your mobile device (such as a phone or tablet) to your Alexa device.

Note: Echo Dot also supports A2DP SRC. This profile allows you to stream audio from Echo Dot to a Bluetooth speaker.

- Audio / Video Remote Control Profile (AVRCP)

This profile allows you to use hands-free voice control when a   mobile device is connected to your Alexa device.

Check the batteries on your Bluetooth device
- If your Bluetooth device has a non-removable battery, make sure the device has a full charge.
- If your Bluetooth device has replaceable batteries, try new batteries.
- Check for interference
- Move your Bluetooth device and Alexa device

away from sources of possible interference, such as microwave ovens, baby monitors, and other wireless devices.
- Make sure your Bluetooth device is close to your Alexa device when you pair it.
- Clear all Bluetooth devices
- Echo and Amazon Tap
- Open the Alexa app
- Open the left navigation panel, and then select Settings.
- Select your Alexa device.
- Select Bluetooth > Clear.
- Echo Dot
- Open the Alexa app.
- Open the left navigation panel, and then select Settings.
- Select your Alexa device.
- Select Bluetooth.
- Select a device from the list, and then select Forget. Repeat this step for all other Bluetooth devices in the list.
- After you clear all devices, restart your Alexa device and your Bluetooth device.
- Pair your Bluetooth devices again
- Echo and Amazon Tap
- From your mobile device, open the settings menu and turn on Bluetooth. Make sure you're near your Alexa device when you do this.
- Say, "Pair." Your Alexa device enters pairing mode.
- From the Bluetooth settings menu on your mobile device, select your Alexa device. Alexa then tells you if the connection is successful.

- Echo Dot
- Open the Alexa app.
- Open the left navigation panel, and then select Settings.
- Select your Alexa device, and then select Bluetooth > Pair a New Device. Your Alexa device enters pairing mode.
- Open the Bluetooth settings menu on your mobile device, and select your Alexa device. Alexa then tells you if the connection is successful.

When the Alexa App Doesn't Work

If you're unable to open the Alexa app, or you receive an error message (such as "The Alexa app is offline"), here are some common solutions that may help.

Compatibility

- iOS 7.0 or higher
- Fire OS 2.0 or higher
- Android 4.4 or higher

Computer web browsers

- Firefox
- Microsoft Edge
- Chrome
- Safari
- Internet Explorer (10 or higher)

iOS

- Restart your iPhone, iPad, or iPod touch
- Press and hold the Sleep / Wake button on your iOS device until a slider appear or until it restarts itself.
- Press and drag the slider to turn the device off.
- After your device turns off, press and hold the Sleep / Wake button to restart the device.
- Force close the app
- Press the Home button twice until previews of

your most recently used apps appear on the screen.
- Swipe until you locate the Alexa app, and then swipe up on the app to close it.
- Uninstall and reinstall the app
- Press and hold the Alexa app until it "shakes" on the screen, and then tap the X on the app.
- After you uninstall the app, go to the Apple App Store and install the Alexa app again.
- Android
- Restart your Android device
- Press and hold the Power button, and then select the option to turn off your device.
- After your device turns off, press the Power button again to restart the device.
- Force close the app
- From the Home screen of your Android device, go to Settings > Apps (or Applications). For some devices, you may need to select Manage applications next.
- Locate Alexa from your list of installed apps, and then select Clear Data.
- After you clear the application data, select Force Stop.
- Uninstall and reinstall the app
- From the app menu on your Android device, select Alexa, and then select Uninstall.
- After you uninstall the app, go to Google Play and install the Alexa app again.

Fire OS

- Restart your Fire OS device

- Press and hold the Power button, and then select the option to turn off your device.
- After your device turns off, press the Power button again to restart the device.
- Force close the app
- From the Home screen of your Fire OS device, swipe down from the top of the screen to open Quick Settings, and then tap Settings OR More > Apps & Games OR Applications > Manage All Applications OR Installed Applications.
- Locate Alexa from your list of installed apps, and then select Clear Data.
- After you clear the application data, select Force Stop.
- Uninstall and reinstall the app
- From the app menu on your device, select Alexa, and then select Uninstall.
- After you uninstall the app, go to the Apps library, and then select Alexa to download the app again.
- Web Browser

- Reload the web page.
- Clear the cache and cookies from your browser. When you clear the cache and cookies, this removes any website settings (such as usernames and passwords) from your browser.
- Close your web browser and open it again.

Streaming Issues

If you experience issues streaming music, audiobooks, and more on your Echo, Echo Dot or Amazon Tap device, here are some common solutions that may help.

A low internet connection or low bandwidth is the common cause of having streaming issues with your devices. To stream music, audiobooks, and other content through Alexa, your Internet connection needs to be at a speed of at least 512 Kbps (0.51 Mbps).

Reduce Wi-Fi congestion
If you have multiple devices on your Wi-Fi network, you may have inconsistent Wi-Fi performance.

- Turn off devices you aren't using to free up bandwidth on your network.
- Move your Alexa device closer to your router and modem if it's in a different room or blocked by an object.
- Move your Alexa device away from walls, metal objects, and other sources of possible interference (such as microwave ovens or baby monitors).
- If your Alexa device is on the floor, move it to a higher location.
- (Echo and Echo Dot) Connect to your router's 5 GHz channel
- Many Wi-Fi devices only connect to the 2 GHz channel. If multiple devices use this channel on your network, your network speed may be

slower. If you have a dual-band router, you can connect your Echo device to the less congested 5 GHz channel for better range and less interference.

Restart your Alexa device and network hardware
- You can restart your Alexa device, Internet modem, and/or router to resolve most intermittent Wi-Fi issues.
- Turn off your router and modem, and then wait 30 seconds.
- Turn on your modem, and then wait for it to restart.
- After you restart your modem, turn on your router, and then wait for it to restart.
- After you restart your network hardware, turn your Alexa device off and on again.

If you're still unable to stream any content on your Alexa device, contact your Internet service provider, router manufacturer, or network administrator for help.

Software Updates

Your Alexa device receives software updates automatically over Wi-Fi. These updates usually improve performance and add new Alexa features so be sure to update them if they don't update it automatically.

Echo Devices

- Latest Software Version: 3417
- Bug fixes and performance improvements.
- New and Enhanced Features

Amazon Tap

- Latest Software Version: 150003820
- New and Enhanced Features
- Bug fixes and performance improvements.

For information about Fire TV devices, go to **"Kindle/Fire Software Updates"**.

To determine the current software version for your Alexa device:

- Open the Alexa app.
- Open the left navigation panel, and then select Settings.

- Select your device, and then scroll down until you see Device software version.

To download the latest software update for your Alexa device:

- Make sure your device is on and has an active Wi-Fi connection.
- Avoid saying anything to your device.
- When the update is ready to install, the light ring on your device will turn blue and then the device will install the latest update. Keep in mind that it may take up to 15-20 minutes to install a software update, depending on your Wi-Fi connection.

Note: Having trouble updating your Alexa device? Restart your device first to see if it resolves your problem.

Kindle/Fire Software Updates

New software updates become available for certain Kindle e-readers, which are needed to continue downloading Kindle content, access to the store and to use services like Whispersync, Goodreads and Kindle's Freetime. Whenever there's an update, it should automatically install once your device is charged and connected to the internet.

Download a Software Update Wirelessly

- Plug your Kindle in to charge during the update.
- Connect your Kindle to a Wi-Fi network. The update should download and begin automatically, even if your Kindle is in sleep mode.
- Leave your Kindle connected to both power and Wi-Fi until the update is complete. To check if the update successfully took place, check your Library and sort by **Recent** for a confirmation letter titled **07-2016 Kindle Software Update** (for Kindle Keyboard 3rd Generation, Kindle Touch 4th Generation, or Kindle 5th Generation) or **2015 Your Kindle is Updated** (for Kindle Paperwhite 5th Generation or Kindle Paperwhite 6th Generation).
- Your Kindle may restart multiple times and you may also see "Your Kindle is Updating" on the screen. Once the update is complete, your Kindle should automatically restart and you should once again have access to all Kindle services on your device.

The Amazon Echo is really a piece of cake once you get used to it. Hopefully you don't have too much technical difficulties while using it but if you do, refer to the many sections above to help you out. I personally think it's an outstanding device and that nothing else beats it. It's fun, convenient and informational. Have fun using your Echo and getting to know Alexa, I know I did when I first purchased it! Alexa will definitely make your daily life easier and also grows even better within time and updates! I highly recommend this product to anyone that's looking for a speaker device that can do about anything!

www.ingramcontent.com/pod-product-compliance
Lightning Source LLC
Chambersburg PA
CBHW070334190526
45169CB00005B/1884